Common Core Standards

Algebra I Practice Tests

Common Core State Standards ® Introduction & Curriculum

The Common Core State Standards provide a consistent, clear curriculum about what students are expected to learn, so teachers and parents know what they need to do to help. Various states and school districts offer tests that measure student proficiency. These practice tests are based on the Common Core Standards curriculum at http://www.corestandards.org/

Contents

Algebra I Questions: Section 1 4

Algebra I Questions: Section 2 35

Algebra I Solutions: Section 1 66

Algebra I Solutions: Section 2 87

Algebra I Questions: Section 1

1. In the equation, $1(x + y) = x + y$, what property of whole numbers is being illustrated?

    a. No property is being demonstrated in this problem.

    b. The Distributive Property of Multiplication over Subtraction.

    c. The Associative Property for Multiplication.

    d. The Identity Property for Multiplication.

2. What is the multiplicative inverse of $-\frac{1}{3}$?

    a. 3

    b. -3

    c. 0.33

    d. $\frac{1}{3}$

3. $\sqrt{36} + \sqrt[3]{64} =$

    a. 14

    b. 17

    c. 10

    d. 6

4. What is $\dfrac{x^2-8x+12}{x^2-3x+2}$ reduced to lowest terms?

a. $\dfrac{x-6}{x-1}$

b. $\dfrac{x-2}{x+2}$

c. $\dfrac{x-6}{x-2}$

d. $\dfrac{x+2}{x-1}$

5. Which of the following statements is a valid conclusion to the statement "If a high school student is an art club member, then the student is a good artist."

a. All students are good artists.

b. All high school art club members are good artists.

c. A student attending high school is in art club.

d. All good artists are high school art club members.

6. Which *best* represents the graph of $y = x - x^2$?

a.

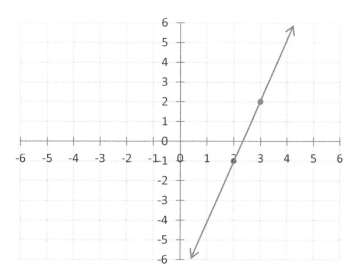

b.

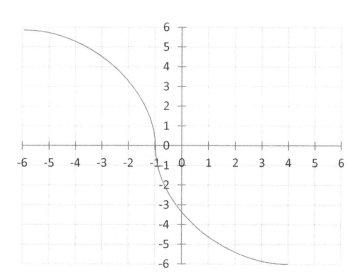

c.

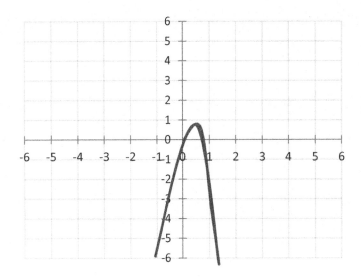

d.

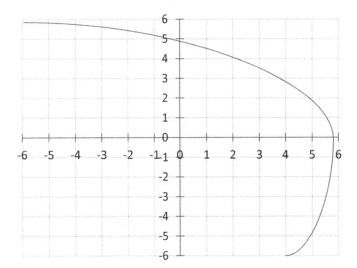

7. What is the domain of the function showed below?

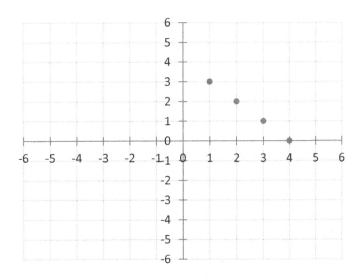

a. {0,1,2,3}
b. {1,2,3,4}
c. {-1,-2,-3,-4}
d. {0, -1,-2,-3}

8. What is the y-intercept of the graph $3x + 4y = 20$?

a. 5

b. 6

c. 4

d. 10

9. Which equation represents a line that is parallel to $y = -\frac{1}{3}x + 1$?

a. $y = -3x + 1$

b. $y = \frac{1}{3}x + 5$

c. $y = 3x + 5$

d. $y = -\frac{1}{3}x + 5$

10. Which ordered pair is the solution to the system of equations below?

$$\begin{cases} 3x + y = 15 \\ 6x - y = 12 \end{cases}$$

a. (-4,1)
b. (6,3)
c. (3,6)
d. (1,12)

11. What expression is equivalent to $x^4 x^5$?

a. $x^3 x^5$

b. $x^3 x^2$

c. $x^3 x^4$

d. $x^3 x^6$

12. Which relation is a function?

    a. {(3,-1), (3,-2), (3,-3), (3,-4)}

    b. {(7,7), (8,7), (8,9), (11,9)}

    c. {(1,7), (2,6), (3,5), (4,4)}

    d. {(4,2), (5,3), (4,6), (7,8)}

13. Michael will make a 34% acid solution during chemistry lab. He has already poured a 3 fl. oz. of a 70% acid solution into a beaker. How many fl. oz. of an 8% acid solution must he add to this to create the desired mixture?

    a. 4.15 fl. oz.

    b. 15 fl. oz.

    c. 3.15 fl. oz.

    d. 4.5 fl. oz.

14. What quantity should be added of this equation to complete the square?

$$x^2 + 16x = 36$$

    a. 32
    b. 64
    c. -64
    d. -32

15. What is the solution to the inequality $x - 7 < 9$?

    a. $x > 2$

    b. $x < 16$

    c. $x < 2$

    d. $x > 16$

16. Which is a factor of $x^2 - 18x + 32$?

    a. $x + 2$

    b. $x - 4$

    c. $x - 8$

    d. $x - 2$

17. What is the complete factorization of $9a^2 - 81$?

    a. $9(a^2 - 81)$

    b. $9(a - 9)^2$

    c. $-9(a - 9)^2$

    d. $9(a - 3)(a + 3)$

18. Heather correctly solved the equation $x^2 + 16x = 4$ by completing the square. Which equation is part of her solution?

   a. $(x+8)^2 = 20$

   b. $(x+8)^2 = 8$

   c. $(x+4)^2 = 20$

   d. $(x+4)^2 = 16$

19. What inequality does the shaded region of the graph represent?

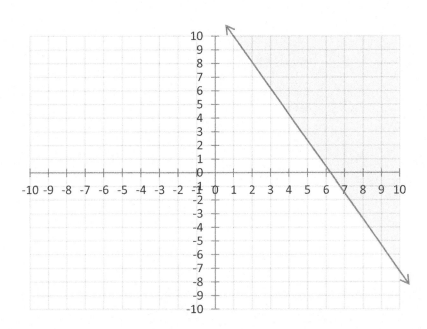

   a. $3x + y \geq -8$
   b. $3x + y \leq -8$
   c. $3x + y \geq 8$
   d. $3x + y \leq 8$

20. Which graph is a function of x?

a.

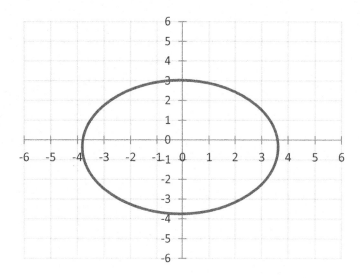

b.

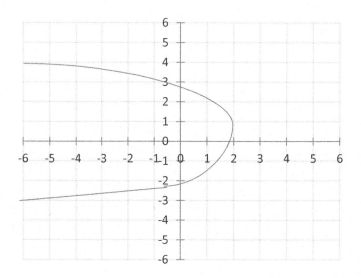

c.

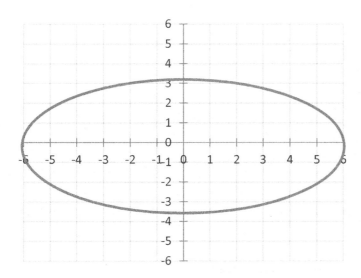

d.

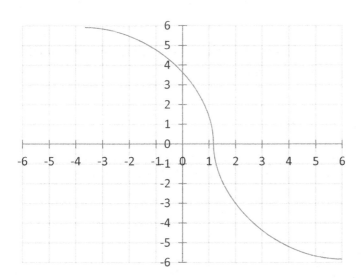

21. What is the equation of the line has a slope of 3 and passes through the point of (2, -6)?

a. $y = 3x + 9$

b. $y = 3x - 9$

c. $y = 3x - 12$

d. $y = 3x + 12$

22. The data in the table shows the cost of renting a skateboard by the hour, including a deposit.

Renting a Skateboard

Hours(h)	Cost in dollars (c)
2	12
5	24
8	36

If hours, h, were graphed on the horizontal axis and the cost, c, were graphed on a vertical axis, what would be the equation of a line that fits the data?

a. $c = 4h - 4$
b. $c = 4h + 4$
c. $c = 4h$
d. $c = \frac{1}{4}h + 4$

23. What is the solution for this equation?

$|2x - 4| = 14$

a. $x = -5$ or $x = 9$
b. $x = -5$ or $x = 2$
c. $x = 2$ or $x = 9$
d. $x = 5$ or $x = -9$

24. What is the reciprocal of $-\frac{1}{4}$?

    a. 4

    b. 0.25

    c. $\frac{1}{4}$

    d. -4

25. What is the solution set for the inequality?

$$-7|2 - x| < -21$$

    a. x < -2 or x > 5
    b. x < -5 or x > -2
    c. x < -1 or x > 5
    d. x < -1 or x > -5

26. Which equation is equivalent to the inequality $7x - 3(4x + 2) = 10x$?

    a. $-5x + 6 = 10x$

    b. $-5x - 1 = 10x$

    c. $-5x - 6 = 10x$

    d. $19x - 6 = 10x$

27. A 175-inch-long piece of string is cut into 3 pieces. The first piece is twice as long as the second piece of string. The third piece is four times as long as the second piece of string. What is the length of the longest piece of string?

   a. 125 inches

   b. 100 inches

   c. 50 inches

   d. 25 inches

28. The lengths of a side of a triangle are $y, y + 2$, and 8 centimeters. If the perimeter is 64 centimeters, what is the value of y?

   a. 27

   b. 24

   c. 54

   d. 55

29. Which number serves as a counterexample to the statement below?

   | All multiples of 9 have digits that will add up to 9. |

   a. None, all of these numbers will add up to 9.
   b. 135
   c. 180
   d. 162

30. Dawn's solution to an equation is shown below.

Given: $3(x + 2) = 12$

Step 1: $3 * x + 3 * 2 = 12$

Step 2: $3x + 6 = 12$

Step 3: $3x = 6$

Step 4: $x = 2$

Which property of real numbers did Dawn use for Step 1?

    a. zero product property of multiplication
    b. distributive property of multiplication over subtraction
    c. distributive property of multiplication over addition
    d. commutative property of multiplication

31. Frank's solution to an equation is shown below.

Given: $z + 7(z - 4) = 20$

Step 1: $z + 7z - 28 = 20$

Step 2: $8z - 28 = 20$

Step 3: $8z = 20 - 28$

Step 4: $8z = -8$

Step 5: $\dfrac{8z}{8} = \dfrac{-8}{8}$

Step 6: $z = -1$

What statement about Frank's solution is true?

    a. Frank's solution is correct.
    b. Frank made a mistake in Step 2.
    c. Frank made a mistake in Step 3.
    d. Frank made a mistake in Step 5.

32. When is this statement true?

> When a number is squared it remains the original number.

a. This statement is never true.
b. This statement is true for the numbers zero and one.
c. The statement is always true.
d. The statement is always true for the number five.

33. The chart below shows an expression evaluated for four different values of x.

x	$x^2 - 5x + 10$
1	6
2	4
6	16
8	34

Jordan concluded that for all positive values of x, $x^2 - 5x + 10$ produces a number ending in 4 or 6. Which value of x serves as a counterexample to prove Jordan's conclusion false?

    a. 3
    b. 4
    c. 7
    d. 10

34. Which *best* represents the graph of $y = 4x + 1$?

a.

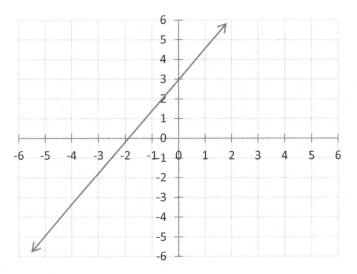

b.

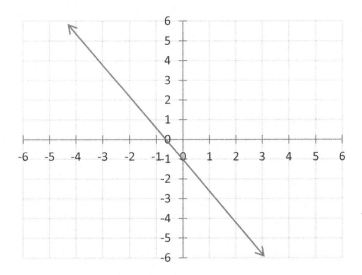

c.

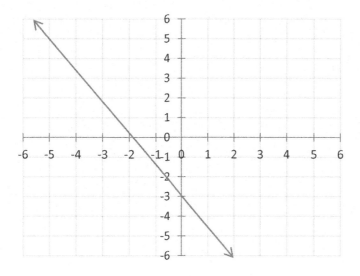

d.

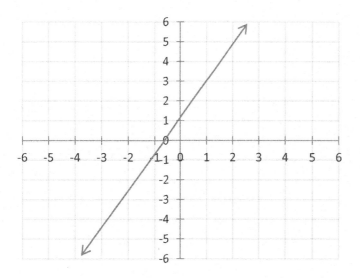

35. Which inequality is shown on the graph below?

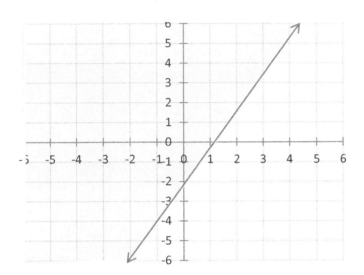

a. $y > \frac{1}{3}x - 2$

b. $y < \frac{1}{3}x - 2$

c. $y \geq \frac{1}{3}x - 2$

d. $y \leq \frac{1}{3}x - 2$

36.

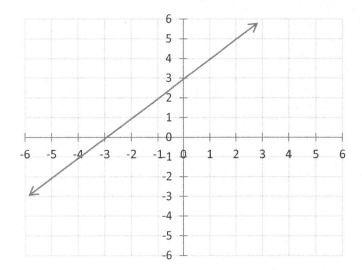

Which equation *best* represents the graph above?

    a. $y = x + 3$
    b. $y = x - 3$
    c. $y = 3x$
    d. $y = x$

37. Which point lies on the line defined by $2x + 4y = 32$?

    a. (0,6)

    b. (6,5)

    c. (0,-5)

    d. (5,6)

38. What is the solution to this system of equations?

$$\begin{cases} y = -2x + 10 \\ 3x + 4y = 5 \end{cases}$$

a. no solution
b. infinitely many solutions
c. (1,8)
d. (7,-4)

39. Which of the following *best* describes the graph of this system of equations?

$$\begin{cases} -4x = 3y + 5 \\ 8x = 6y - 1 \end{cases}$$

a. two lines intersecting at one point
b. two lines intersecting at two points
c. two parallel lines
d. two identical lines

40. Andrea has a total of 100 dimes and quarters. If the total value of coins is $15.85, how many quarters does she have?

a. 61

b. 52

c. 34

d. 39

41. The equation of line l is $y = 4x + 6$, and the equation of line q is $-y = -4x + 7$. Which statement about these two lines is true?

    a. Lines l and q are parallel.

    b. Lines l and q are perpendicular.

    c. Lines l and q have the same x-intercept.

    d. Lines l and q have the same y-intercept.

42. $(3x^2 + 2x - 6) - (x^2 - x + 2)$

    a. $4x^2 + 3x + 8$

    b. $4x^2 + x - 4$

    c. $2x^2 + 3x - 8$

    d. $2x^2 + x + 4$

43. The sum of two binomials is $6x^2 + 15$. If one of the binomials is $-2x^2 + 9$, what is the other binomial?

    a. $4x^2 + 8$

    b. $8x^2 + 6$

    c. $4x^2 + 6$

    d. $8x^2 + 8$

44. Which of the following shows $5z^2 + 21z - 54$ completely factored?

a. $(5z - 2)(z + 27)$

b. $(5z - 9)(z + 6)$

c. $(5z + 2)(z - 27)$

d. $(5z + 9)(z - 6)$

45. What is the factored form of $4a^2 - 28ab + 40b^2$?

a. $(4a - 2b)(a - 5b)$

b. $4(a - 8b)(a - 5b)$

c. $(2a - 4b)(2a - 4b)$

d. $4(a - 2b)(a - 5b)$

46. $\dfrac{4x^5}{8x^8}$

a. $\dfrac{1}{2x^3}$

b. $2x^3$

c. $\dfrac{x^2}{4}$

d. $\dfrac{1}{4x^2}$

47. Which of the following expressions is equal to $(x+3)+(x+3)(3x-1)$?

    a. $4x^2 - 9x$

    b. $3x^2 - 10x$

    c. $3x^2 + 9x$

    d. $4x^2 + 10x$

48. If x is subtracted from x^2, the answer is 20. Which of the following could be the value of x?

    a. -8

    b. -4

    c. 15

    d. 10

49. Gavin is solving this equation by factoring.

$$6x^2 - 21x + 18$$

Which could be one of his correct factors?

    a. $3x + 6$
    b. $3x - 6$
    c. $x + 6$
    d. $x - 6$

50. Kyle has started to solve this equation by completing the square but has forgotten one of the steps along the way.

$$ax^2 + bx + c = 0$$

Step 1: $ax^2 + bx = -c$

Step 2: ?

Step 3: $x^2 + \frac{b}{a}x + \left(\frac{b}{2a}\right)^2$

What should be Step 2 in the solution?

   a. $x + \frac{b}{a} = -\frac{c}{ax}$
   b. $x^2 = \frac{b}{2a} - \frac{c}{a}$
   c. $\left(x + \frac{b}{2a}\right)^2 = \frac{b^2 - 4ac}{4a^2}$
   d. $x^2 + \frac{b}{a}x = -\frac{c}{ax}$

51. Taylor is writing down the first five steps to derive the quadratic formula shown below.

> Step 1: $ax^2 + bx = -c$
> Step 2: $x^2 + \frac{b}{a}x = -\frac{c}{ax}$
> Step 3: $\left(x + \frac{b}{2a}\right)^2 = \frac{b^2 - 4ac}{4a^2}$
> Step 4: $x^2 + \frac{b}{a}x + \left(\frac{b}{2a}\right)^2$
> Step 5: $x = \pm\sqrt{\frac{b^2 - 4ac}{4a^2}} - \frac{b}{2a}$

The first two steps are in the correct order but the last three are not. What is the correct order for the last three steps.

   a. 4,3,5
   b. 3,4,5
   c. 5,3,4
   d. 5,4,3

52. What is the solution set of the quadratic equation $9x^2 + 12x + 4 = 0$?

    a. no real solution

    b. $\left\{\frac{1}{2}, -1\right\}$

    c. $\left\{-\frac{4}{3}\right\}$

    d. $\left\{-\frac{2}{3}\right\}$

53. Which statement *best* explains why there is no real solution to the quadratic equation $3x^2 + x + 8 = 0$?

    a. The value of $1^2 - 4(3)(8)$ is not a perfect square.

    b. The value of $1^2 - 4(3)(8)$ is negative.

    c. The value of $1^2 - 4(3)(8)$ is positive.

    d. The value of $1^2 - 4(3)(8)$ is equal to zero.

54. Which is one of the solutions to the equation $7x^2 - x - 6 = 0$?

    a. $x = \frac{-1+\sqrt{169}}{14}$

    b. $x = \frac{-1}{14} + \sqrt{169}$

    c. $x = \frac{1-\sqrt{169}}{14}$

    d. $x = \frac{1}{14} - \sqrt{169}$

55. Which quadratic function, when graphed, x-intercept, has intercepts of -3 and -4?

a. $y = (x - 3)(x - 1)$

b. $y = (x + 3)(x + 4)$

c. $y = (x + 4)(x + 1)$

d. $y = (x + 3)(x - 4)$

56. How many times does the graph $y = x^2 - 8x + 16$ intersect the x-axis?

a. zero

b. one

c. two

d. three

57. The height of a triangle is 4 inches greater than twice its base. The area of the triangle is 224 square inches. What is the base of the triangle?

a. 4 inches

b. 10 inches

c. 14 inches

d. 16 inches

58. Simplify $\frac{4x^2+22x+30}{2x^2-3x-5}$ to lowest terms.

a. $\frac{2(x+3)}{x-3}$

b. $\frac{2(x+3)}{2x-5}$

c. $\frac{2(x+5)}{x-3}$

d. $\frac{2(x+3)}{2x+5}$

59. What is $\frac{3n^3-2n^2}{9a^2-4}$ reduced to lowest terms?

a. $\frac{n(3n^2+2)}{3n-2}$

b. $\frac{n^2}{3n+2}$

c. $\frac{3n(n^2-2)}{3n+2}$

d. $\frac{n^2}{3n-2}$

60. $\frac{x^2-4}{x+5} \div \frac{3x-6}{x^2+6x+5} =$

a. $\frac{(x+2)(x+1)}{3}$

b. $\frac{x+1}{3(x-2)}$

c. $\frac{(x+2)(x+1)}{x-2}$

d. $\frac{(x-2)(x+1)}{3}$

32

61. Which fraction equals the product $\left(\frac{x+4}{2x-1}\right)\left(\frac{2x-1}{3x-4}\right)$?

    a. $\frac{2x^2+7x-4}{6x^2-11x+4}$

    b. 3

    c. $\frac{2x-1}{3x-4}$

    d. $\frac{x+4}{3x-4}$

62. $\frac{8}{x+3} \cdot \frac{2x+10}{4} =$

    a. $\frac{16x+80}{4x+12}$

    b. $\frac{x+10}{x+3}$

    c. $\frac{4x+10}{4x+3}$

    d. $\frac{4x+40}{x+3}$

63. Two cars leave Fresno, California at the same time in opposite directions. One car has a speed of 70 mph and the other car's speed is 40 mph. In about how many hours will the two cars be 330 miles apart?

    a. 4.5

    b. 3

    c. 6.25

    d. 5

64. Mandy's average driving speed for a 5 hour driving trip was 55 miles per hour. During the first three hours of her trip, she drove 45 miles per hour. What was her average driving speed for the last two hours?

    a. 60 miles per hour

    b. 65 miles per hour

    c. 70 miles per hour

    d. 75 miles per hour

65. What is $\dfrac{x^2+9y^2}{x^2-5xy+6y^2}$ reduced to lowest terms?

    a. $\dfrac{(x+3y)}{(x-2y)}$

    b. $\dfrac{(x+3y)}{(x-3y)}$

    c. $\dfrac{(x-3y)}{(x-2y)}$

    d. $\dfrac{(x-2y)}{(x-3y)}$

Algebra I Questions: Section 2

1. In the equation, $6(x + 2) = (x + 2)6$, what property of whole numbers is being illustrated?

    a. The Distributive Property of Multiplication over Addition.

    b. The Commutative Property of Multiplication.

    c. The Identity Property for Multiplication.

    d. No property is being demonstrated in this problem.

2. $\sqrt[3]{27} + \sqrt{81} =$

    a. 18

    b. 21

    c. 6

    d. 12

3. Which expression is equivalent to $x^3 x^7$?

    a. $x^4 x^7$

    b. $x^3 x^6$

    c. $x^6 x^4$

    d. $x^2 x^7$

4. What is the reciprocal of $\frac{3}{4}$?

    a. $\frac{4}{3}$

    b. 1

    c. $-\frac{3}{4}$

    d. 0.75

5. What is the multiplicative inverse of -8?

    a. $-\frac{1}{8}$

    b. 8

    c. $-\frac{8}{1}$

    d. $\frac{1}{8}$

6. What is the solution for this equation?

$$|3x + 6| = 9$$

    a. $x = 1 \ or \ x = -5$
    b. $x = -1 \ or \ x = 3$
    c. $x = -1 \ or \ x = 5$
    d. $x = 1 \ or \ x = -3$

7. What is the solution to this equation?

$$-2|2 - x| = 12$$

a. $x = 4$ or $x = -8$
b. $x = 2$ or $x = -4$
c. $x = -2$ or $x = 8$
d. $x = -8$ or $x = 8$

8. Which equation is equivalent to $-2x + 3(-5 - x) = x$?

a. $-x + 15 = x$

b. $-5x - 15 = x$

c. $2x - 2 = x$

d. $5x + 2 = x$

9. Which equation is equivalent to $7(x - 3) - 6x + 4 = 2 - (x + 3)$?

a. $7x = 16$

b. $2x = 30$

c. $2x = 16$

d. $7x = 30$

10. Solve: $3(4x - 6) = -x + 34$

Step 1: $12x - 18 = -x + 34$

Step 2: $13x - 18 = 34$

Step 3: $13x = 16$

Step 4: $x = \frac{16}{13}$

Which is the first *incorrect* step in the solution shown above?

    a. Step 1
    b. Step 2
    c. Step 3
    d. Step 4

11. What is the solution to the inequality $x - 4 > 7$?

    a. $x < 3$

    b. $x < 11$

    c. $x > 3$

    d. $x > 11$

12. Which serves as a counterexample to the statement below.

> All numbers ending in 7 are prime numbers.

    a. 7
    b. 17
    c. 27
    d. 37

13. What is the conclusion of the statement in the box below?

$$\text{If } x^2 = 9, \text{ then } x = -3 \text{ or } x = 3.$$

a. $x = -3$ or $x = 3$
b. $x = 3$
c. $x = -3$
d. $x^2 = 9$

14. The chart below shows an expression evaluated for four different values of x.

x	$x^2 + 3x - 7$
1	-3
2	3
5	33
7	63

Taylor concluded that for all positive values of x, $x^2 + 3x - 7$ produces a number ending in 3. Which value of x serves as a counterexample to prove Taylor's conclusion false?

a. 3
b. -4
c. 10
d. 12

15. Janelle's solution to an equation is shown below.

Given: $\frac{x}{5} = 6$

Step 1: $\frac{x}{5}(5) = 6(5)$

Step 2: $x = 30$

Which property of real numbers did Janelle use for step 1?

a. commutative property of multiplication
b. distributive property of multiplication
c. zero product property of multiplication
d. multiplication property of equality

16. Isaac's solution to an equation is shown below.

Given: $n + 4(n + 10) = 80$

Step 1: $n + 4n + 10 = 80$

Step 2: $5n + 10 = 80$

Step 3: $5n = 80 - 10$

Step 4: $5n = 70$

Step 5: $\frac{5n}{5} = \frac{70}{5}$

Step 6: $n = 14$

Which statement is true about Isaac's solution is true?

　　a. Isaac's solution is correct.
　　b. Isaac made a mistake in Step 1.
　　c. Isaac made a mistake in Step 2.
　　d. Isaac made a mistake in Step 3.

17. When is this statement true?

> Every number has a reciprocal except zero.

　　a. This statement is never true.
　　b. This statement also applies to negative numbers.
　　c. This statement also applies to the number one.
　　d. This statement is always true.

18. What is the y-intercept of the graph $5x - 4y = -8$?

　　a. 2

　　b. -4

　　c. -1

　　d. 6

19. Which inequality is shown on the graph below?

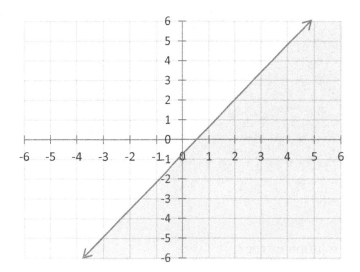

a. 3y > x+2
b. 3y < x-2
c. 3y > 4x-2
d. 3y < 4x+2

20. Which best represents the line $y = -\frac{1}{2}x + 3$?

a.

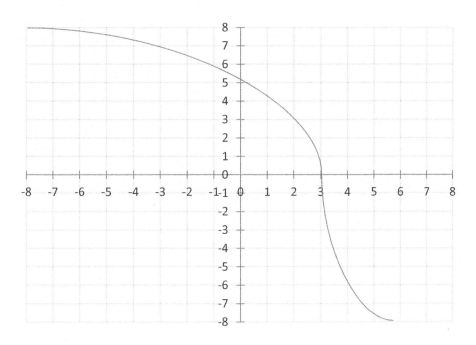

b.

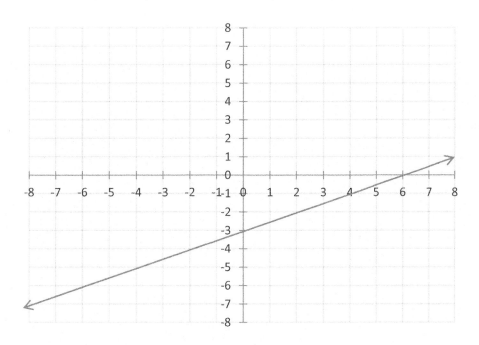

c.

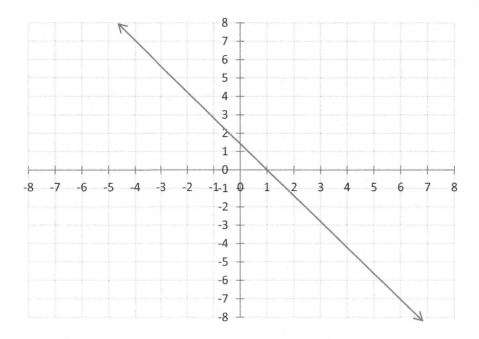

d.

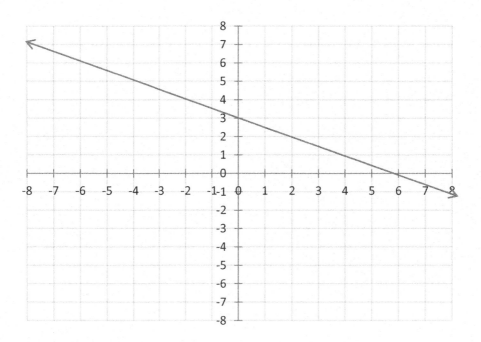

21. Which inequality does the shaded region of the graph represent?

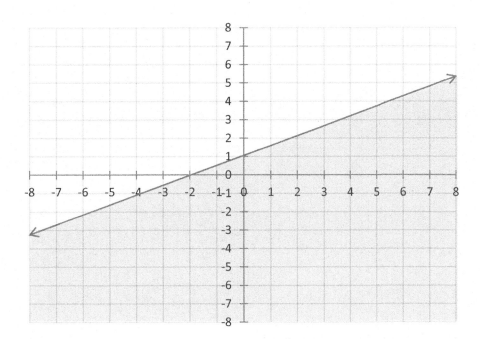

a. $y \geq \frac{1}{2}x + 1$
b. $y \leq \frac{1}{2}x + 1$
c. $y \geq \frac{1}{2}x - 1$
d. $y \leq \frac{1}{2}x - 1$

22.

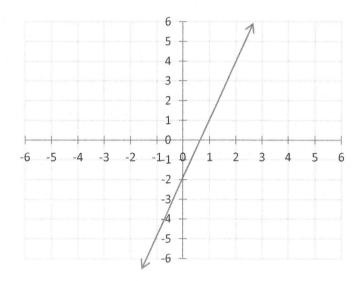

Which equation *best* represents the graph above?

   a. $y = x$
   b. $y = 3x$
   c. $y = 3x - 2$
   d. $y = 3x + 2$

23. Which point lies on the line defined by $2x - 3y = 3$?

   a. $(0,1)$

   b. $(0,3)$

   c. $\left(1, -\dfrac{1}{9}\right)$

   d. $\left(1, -\dfrac{1}{3}\right)$

24. What is the equation of the line that has a slope of -4 and passes through the point of (5,-6)?

    a. $y = -4x + 14$

    b. $y = -4x - 14$

    c. $y = -4x + 21$

    d. $y = -4x - 21$

25. Some ordered pairs for a linear function of x are given in the table below.

x	y
1	15
3	23
5	31
7	39

Which of the following equations was used to generate the table above?

    a. $y = 2x + 15$
    b. $y = 4x + 11$
    c. $y = 4x + 14$
    d. $y = 2x + 13$

26. Which equation represents a line that is parallel to $y = \frac{2}{3}x + 4$?

    a. $y = \frac{3}{2}x + 4$

    b. $y = -\frac{2}{3}x + 3$

    c. $y = \frac{2}{3}x + 2$

    d. $y = -\frac{3}{2}x + 1$

27. The equation of line a is $-\frac{3}{2}x + y = -1$, and the equation of line b is $2x - 3y = 3$. Which statement about the two lines is true?

   a. Lines a and b have the same y-intercept.

   b. Lines a and b are parallel.

   c. Lines a and b have the same x-intercept.

   d. Lines a and b are perpendicular.

28. Which of the following *best* describes the graph of this system of equations?

$$\begin{cases} y = 2x + 1 \\ y = 2x + 5 \end{cases}$$

   a. two identical lines
   b. two parallel lines
   c. two lines intersecting in only one point
   d. two lines intersecting in only two lines

29. Rex buys a book for $16.50, which is 25% discount off the regular price. What is the regular price of the book?

   a. $22.00

   b. $21.50

   c. $20.00

   d. $22.75

30. Which ordered pair is the solution to the system of equations below?

$$\begin{cases} 3x - y = 8 \\ y = x - 2 \end{cases}$$

a. $\left(\frac{3}{2}, 1\right)$
b. (3,1)
c. (3,17)
d. $\left(1, \frac{1}{2}\right)$

31. What is the solution to this system of equations?

$$\begin{cases} 3x + 2y = 16 \\ y = 2x + 1 \end{cases}$$

a. no solution
b. infinitely many solutions
c. (0,1)
d. (2,5)

32. $\dfrac{3x^4}{9x^9} =$

a. $\dfrac{1}{6x^5}$

b. $3x^5$

c. $\dfrac{1}{3x^5}$

d. $\dfrac{x^5}{6}$

33. $(5x^2 - 4x - 7) - (x^2 - 5x - 4) =$

    a. $4x^2 + x - 3$

    b. $4x^2 - 9x - 11$

    c. $4x^2 + x - 11$

    d. $4x^2 - 9x - 3$

34. The sum of two binomials is $7x^2 - 9x$. If one of the binomials is $2x^2 - 3x$, what is the other binomial?

    a. $5x^2 - 6x$

    b. $9x^2 + 12x$

    c. $9x^2 - 6x$

    d. $5x^2 + 12$

35. Which of the following expressions is equal to $(x + 3) + (x - 1)(3x + 2)$?

    a. $3x^2 + x$

    b. $3x^2 - 2$

    c. $3x^2 - 2x$

    d. $3x^2 + 1$

36. The length of a rectangular window is 6 feet more than its width, w. The area of the window is 42 square feet. Which equation could be used to find the dimensions of the window?

    a. $w^2 - 6w - 42$

    b. $w^2 + 6w - 42$

    c. $w^2 - 6w + 42$

    d. $w^2 + 6w + 42$

37. Which is the factored form of $2a^2 + 2ab - 24b^2$?

    a. $(2a - 3b)(a + 4b)$

    b. $2(a + 6b)(a - 2b)$

    c. $(2a - 4b)(a + 3b)$

    d. $2(a - 3b)(a + 4b)$

38. What is a factor of $x^2 - 13x + 36$?

    a. $x - 3$

    b. $x + 3$

    c. $x - 4$

    d. $x + 4$

39. Which of the following shows $4z^2 - 12z + 9$ factored completely?

    a. $(2z + 3)(2z - 4)$

    b. $(2z - 3)^2$

    c. $4(z - 3)(z - 3)$

    d. $4z^2 - 12z + 9$

40. What is the complete factorization of $36 - 4t^2$?

    a. $4(3 + t)(3 - t)$

    b. $-4(3 + t)(3 - t)$

    c. $-4(3 + t)^2$

    d. $4(3 + t)^2$

41. If x^2 is added to x, the sum is 72. Which of the following could be the value of x?

    a. -8

    b. -9

    c. 11

    d. 72

42. What quantity should be added to both sides of this equation to complete the square?

$$x^2 - 6x = 15$$

    a. 9
    b. -9
    c. 3
    d. -3

43. What are the solutions for the quadratic equation $x^2 + 8x = 20$?

    a. -2,-10

    b. 2,-10

    c. -2,10

    d. 2,10

44. John correctly solved the equation $x^2 + 6x = 12$ by completing the square. Which equation is part of his solution?

    a. $(x + 3)^2 = 12$

    b. $(x + 3)^2 = 21$

    c. $(x + 6)^2 = 12$

    d. $(x + 6)^2 = 21$

45. Mariah is solving this equation by factoring.

$$12x^2 - 8x - 20 = 0$$

Which expression could be one of her correct factors?

- a. $x + 5$
- b. $x - 5$
- c. $3x + 5$
- d. $3x - 5$

46. Nancy is solving the equation by completing the square.

$$ax^2 + bx + c = 0$$

What should be the first step in the solution?

- a. $ax^2 + bx = -c$
- b. $(x + \frac{b}{2a})^2 = \frac{b^2 - 4ac}{4a^2}$
- c. $x^2 = \frac{b}{2a} - \frac{c}{a}$
- d. $x + \frac{b}{a} = -\frac{c}{ax}$

47. What is the solution set of the quadratic equation $x^2 + x + 11 = 0$?

- a. $\left\{\frac{-1+\sqrt{43}}{2}, \frac{-1-\sqrt{43}}{2}\right\}$
- b. $\left\{\frac{1}{2}, -\frac{1}{2}\right\}$
- c. no real solution
- d. $\left\{\frac{-1+\sqrt{44}}{2}, \frac{-1-\sqrt{44}}{2}\right\}$

48. Which is one of the solutions to the equation $x^2 - 7x + 9 = 0$?

a. $\dfrac{-7+\sqrt{13}}{2}$

b. $7 - \sqrt{13}$

c. $-\dfrac{7}{2} + \sqrt{13}$

d. $\dfrac{7-\sqrt{13}}{2}$

49. Which quadratic function, when graphed, has *x*-intercepts of 1 and -2?

a. $y = (2x + 2)(x - 4)$
b. $y = (x + 2)(3x + 3)$
c. $y = (2x - 1)(2x + 6)$
d. $y = (2x - 2)(2x + 4)$

50. The graph of the equation $y = x^2 + 6x + 5$ is shown below.

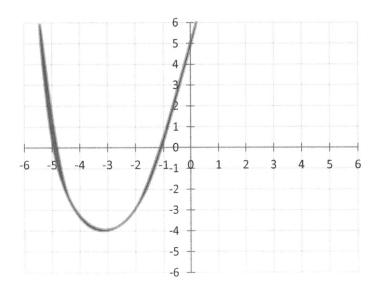

For what value or values of x is $y = 0$?

    a. $x = -1$ and $x = 5$
    b. $x = -1$ and $x = -5$
    c. $x = -1$ only
    d. $x = -5$ only

51. How many times does the graph $y = 3x^2 + 10x + 7$ intersect the x-axis?

    a. none

    b. one

    c. two

    d. three

52. A living room being built measures 11 feet by 15 feet. The owner has now requested it to have a wooden floor trim installed all around it, increasing the total area to 192 square feet. What will be the width of the wooden trim?

    a. 1 ft.

    b. $\frac{1}{2}$ ft.

    c. $\frac{1}{4}$ ft.

    d. 2 ft.

53. The area of a rectangle is 629 square centimeters. The length is 3 more than twice the width. What is the length of the rectangle?

    a. 37 cm

    b. 34 cm

    c. 21 cm

    d. 17 cm

54. What is $\frac{x^2+xy-6y^2}{2xy+6y^2}$ reduced to lowest terms?

    a. $\frac{x+2y}{2}$

    b. $\frac{x+2y}{2y}$

    c. $\frac{x-2y}{2}$

    d. $\frac{x-2y}{2y}$

55. What is $\dfrac{3x^2+4x+1}{3x^2-14x-5}$ reduced to lowest terms?

   a. $\dfrac{x-1}{x+5}$

   b. $\dfrac{x+1}{x-5}$

   c. $\dfrac{x-1}{x-5}$

   d. $\dfrac{x+1}{x+5}$

56. Simplify $\dfrac{2x^2-x-21}{4x^2-49}$ to lowest terms.

   a. $\dfrac{2x+3}{2x+7}$

   b. $\dfrac{x-3}{2x-7}$

   c. $\dfrac{x+3}{2x+7}$

   d. $\dfrac{x-3}{2x+7}$

57. What is $\dfrac{4a^3+8a^2}{a^2-4}$ reduced to lowest terms?

   a. $\dfrac{4a^2}{2(a+1)}$

   b. $\dfrac{a-2}{a+2}$

   c. $\dfrac{a^2}{a-2}$

   d. $\dfrac{4a^2}{a+2}$

58. $\dfrac{5b^2+5}{b^2-3b-4} \cdot \dfrac{b^2-2}{b^2-1}$

    a. $\dfrac{5(b+1)}{b-4}$

    b. $\dfrac{5b(b+1)}{b-4}$

    c. $\dfrac{5(b+1)}{b-1}$

    d. $\dfrac{5b(b+1)}{b-1}$

59. Which fraction equals the product $\left(\dfrac{x-4}{2x-1}\right)\left(\dfrac{3x+2}{x+4}\right)$?

    a. $\dfrac{3x^2+14x-8}{2x^2+8x-4}$

    b. $\dfrac{3x^2-10x-8}{2x^2-7x+4}$

    c. $\dfrac{3x^2-10x-8}{2x^2+7x-4}$

    d. $\dfrac{3x^2+14x+8}{2x^2+7x-4}$

60. $\dfrac{x^2+6x+9}{x+2} \div \dfrac{6x+9}{x^2-4}$

    a. $\dfrac{(x+3)(x-2)}{3}$

    b. $\dfrac{(x+3)(x+2)}{3(x-2)}$

    c. $\dfrac{(x+3)(x+2)}{3}$

    d. $\dfrac{(x+3)^2}{3}$

61. A local café sells a house blend coffee made up of two types of coffee. The first is made from 6 pounds of coffee that sells for $9.40 per pound and the other from 14 pounds of coffee that sells for $8.00 per pound. Find the coffee mixture price per pound.

    a. $8.70

    b. $8.42

    c. $17.40

    d. $11.24

62. Ten grams of sugar are added to a 50g serving of a breakfast cereal that is 20% sugar. What is the percent concentration of sugar in the resulting mixture?

    a. 20%

    b. 25%

    c. 30%

    d. 33%

63. A train left Boston for Nashville travelling at 50 miles per hour. Two hours later another train leaves from Boston to Nashville on the track beside or parallel to the first train but it travels at 75 miles per hour. How far away from Boston will the faster train pass the other train?

    a. 200 miles

    b. 250 miles

    c. 300 miles

    d. 350 miles

64. Which relation is a function?

    a. $\{(1,2),(2,4)(1,6),(3,8)\}$

    b. $\{(0,6),(1,7),(2,8),(3,9)\}$

    c. $\{(0,1),(0,2),(0,4),(0,5)\}$

    d. $\{(3,6),(-3,-6),(3,9),(-3,-9)\}$

65. For which equation graphed below are all the y-values positive?

a.

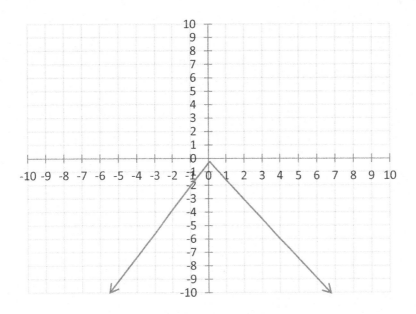

b.

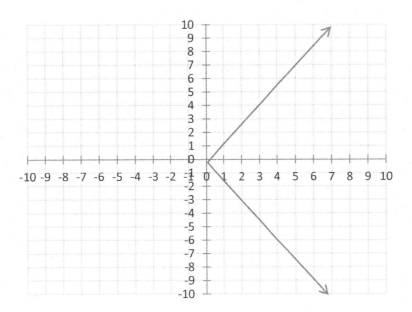

c.

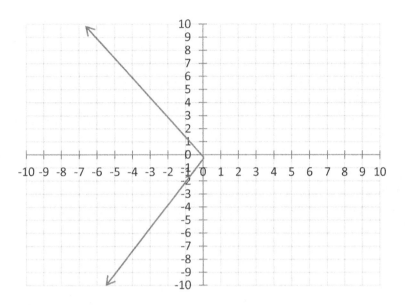

d.

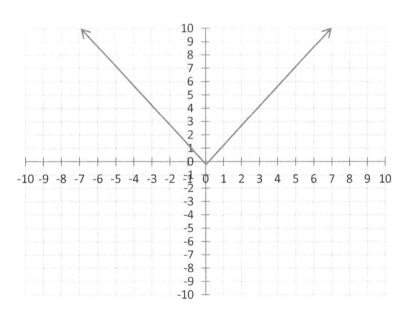

66. Which graph is a *not* function of x?

a.

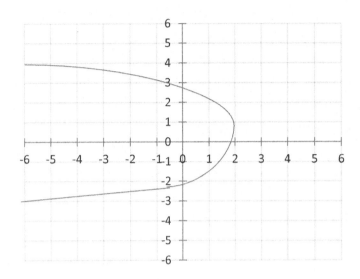

b.

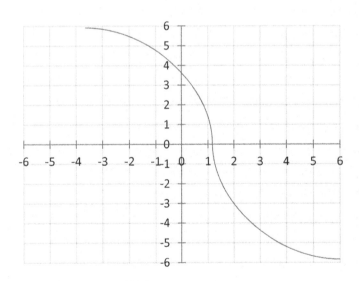

c.

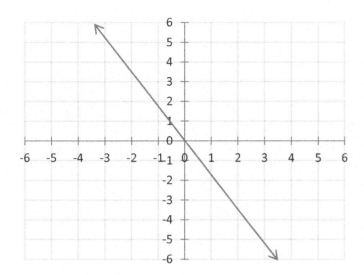

d.

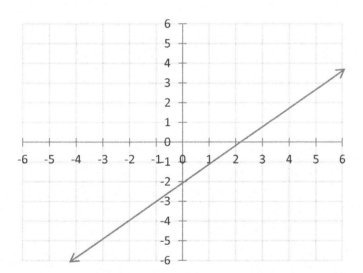

Algebra I Solutions: Section 1

1.

d. The Identity Property for Multiplication.

This equation demonstrates the Identity Property for Multiplication. The number 1 is the unique whole-number such that for every whole number x, $x \cdot 1 = x = 1 \cdot x$.

2.

b. -3

The multiplicative inverse (or reciprocal) is basically a fraction flipped upside down therefore the multiplicative inverse of $-\frac{1}{3}$ is -3.

3.

c. 10

The square root of 36 is 6 and the cubed square root of 64 is 4. Adding together 6 + 4 will give you an answer of 10.

4.

a. $\dfrac{x-6}{x-1}$

The lowest terms for this equation are $\dfrac{x-6}{x-1}$. On the top of the equation after simplifying you are left with x- 6 and x- 2. On the bottom equation after simplifying you are left with x - 2 and x - 1. The x -2 on the top and the x - 2 on the bottom cancel each other out since they are the same leaving you with x- 6 on the top and x - 1 on the bottom.

5.

b. All high school art club members are good artists.

Reading the statement, the most valid conclusion one could make from it was that all high school members are good artists. The other conclusions would not be valid ones to make based upon the statement.

6.

c.

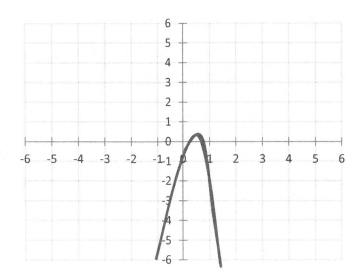

By determining the y-values for a standard set of x-values and writing them in a table:

x	y
-3	-12
-2	-6
-1	-2
0	0
1	0
2	-2
-3	-6

Since $y = 0$ for both $x = 0$ and $x = 1$, what happens in between is $x = \frac{1}{2}$ and $y = \frac{1}{4}$. There are no limits to the x- or y- values with this equation.

7.

b. {1,2,3,4}

When looking for the domain of a function, one is looking for the x-values which in this case are {1,2,3,4} for the problem $y = -x + 4$. The value of y becomes lower as the value of x increases.

8.

a. 5, to find the y-intercept in this equation you must solve for y. To solve for y, you replace x with 0 to find y. $3(0) = 0$ leaving $4y = 20$. Dividing 4 from each side will allow y to be alone and finding it equals 5.

9.

d. $y = -\frac{1}{3}x + 5$

To find a line parallel to the equation $y = -\frac{1}{3}x + 1$, you must remember that for lines to be parallel in a slope they must have the same slope. The slope in the original equation is $-\frac{1}{3}$, therefore the slope in the answer must also be $-\frac{1}{3}$.

10.

c. (3,6)

This is the only ordered pair out of the solutions given that will work in both of the system of equations given.

11.

d. $x^3 x^6$

When multiplying different powers of the same base, you add the exponents. $x^4 x^5 = x^9$ Looking over the expressions given, the only expression that is equivalent to x^9 is $x^3 x^6$.

12.

c. {(1,7), (2,6), (3,5), (4,4)}

C is the only function given in this question. The domain (x) is different from all of the other domains in the other ordered pairs. The domain cannot be the same and be a function. The range (y) can still be the same and still be a function.

13.

a. 4.15 fl. oz.

To solve this problem let x be the number of ounces of the 8% solution that Michael needs to use. By using x, we can create this problem: $0.7(3) + 0.08\,x = 0.34(3 + x)$

Multiply the terms out: $2.1 + 0.08x = 1.02 + 0.34x$

Subtract 1.02 and $0.08x$ from each side: $1.08 = 0.26x$

Divide each side by 0.26: $4.15 = x$

Therefore Michael needs to add 4.15 fl. oz. to make the mixture correctly.

14.

b. 64

You square half the coefficient of x. In this case, $8^2 = 64$. 64 would be added to both sides of the equation to complete the square.

15.

b. $x < 16$

To balance the equation and get x alone, we will add 7 to each side to find the answer.

16.

d. $x - 2$

When factoring the equation $x^2 - 18x + 32$, you get the factors of $x - 2$ and $x - 16$. Therefore one of the factors for the equation is $x - 2$.

17.

d. $9(a - 3)(a + 3)$

To factor the equation completely, first we need to take out the biggest number that is in both $9a^2$ and 81. This number is 9. After removing 9 from the problem we are left with d. $9(a^2 - 9)$. Next we need to factor $a^2 - 9$ to bring it to its simplest form. This would bring us to $9(a - 3)(a + 3)$ allowing the equation to be completely factored.

18.

a. $(x+8)^2 = 20$

Heather had to use the quadratic formula to solve the problem correctly. Since the problem is arranged as $x^2 + ax = b$, we now need to arrange the problem as $(x + c)^2$. When this is multiplied out you get $x^2 + 2cx + c^2$. With the x coefficient being both a and $2c$, c must be $a/2$ you also will get an unwanted c^2. This needs to be taken off to compensate using the equation: $(x + (a/2))^2 = b + a/2^{\,2}$

All that is needed is to figure out the problem using the facts given.

$(x + (\frac{16}{2}))^2 = 4 + (\frac{16}{2})^2$

$(x + 8)^2 = 4 + 4^2$

$(x + 8)^2 = 20$

$x + 8 = \pm\sqrt{20}$

$x = -8 \pm \sqrt{20}$

19.

c. $3x + y \geq 8$

This equation is the only equation out of the four that would match the inequality of the shaded region.

20.

d. Out of the four graphs shown, d is the only one that does not intersect more than at one point causing them not to be functions of x.

21.

c. $y = 3x - 12$

The equation of a line is $y = mx + b$ where m is the slope and b is the value of the point where the line crosses the y-axis.

Since we know the slope is 3, we will replace m with 3 to get: $y = 3x + b$

The line passes through the point $(2, -6)$ so $y = -6$ whenever $x = 2$. We can replace x with 2 and y with 2 to solve for b. The equation now looks like this: $-6 = 3(2) + b$ which can be simplified to: $-6 = 6 + b$.

Now subtract 6 from both sides of the equation to solve for b to get: $b = -12$.

Since we know both m and b, we can write the equation: $y = 3x - 12$

22.

b. $c = 4h + 4$

24 to 12 = 12
36 to 24 = 12

5 to 2 = 3 hours
8 to 5 = 3 hours

For every rate there is an increase in $12 for every 3 hours
12/3 = $4 per hour

2(4) = 8
5(4) = 20
8(4) = 32

Actual costs minus the hourly rate:

12 - 8 = 4
24 - 20 = 4
36 - 32 = 4

Therefore there is an increase in $4 for every rate:

c = 4h + 4

23.

a. $x = -5$ or $x = 9$

Since 2x-4 equal an absolute value, there is more than one possible answer.
Answer 1:
|2x-4| > 14
2x-4 > 14
Add 4 to each side
2x > 18
Divide all sides by 2
x > 9
Answer 2:
2x-4 > -14
Add 4 to each side
2x > -10
Divide by 2
x > -5

24.

d. -4

The reciprocal (or multiplicative inverse) is basically a fraction flipped upside down therefore the reciprocal of $-\frac{1}{4}$ is -4.

25.

c. x < -1 or x > 5

-7|2-x| < -21
Divide 7 from each side: -|2-x| < -3
Then we must do something with the negative: |2-x| > 3 (remember, the sign of the inequality changes when we multiply both sides by -1)

If 2-x is positive then we require
2-x > 3

Subtract 2 from each side: -x > 1
Divide by -1: x < -1

If 2-x is negative then we require
-(2-x) > 3
Multiply by -1: x - 2 > 3
Add 2 to each side: x > 5

Therefore the answer is x < -1 or x > 5

26.

c. $-5x - 6 = 10x$

This equation is equivalent to the original equation given. By using the distributive property, we multiply -3 by $4x$ and 2.

After this is done we are given the equation: $7x - 12x - 6 = 10x$. Next we combine the like terms of $7x$ and $12x$.

Once like terms have been combined, the equation is now $-5x - 6 = 10x$.

27.

b. 100 inches

We will allow x to be the second piece of rope since we know the least about it.

First piece= $2x$
Second piece= x
Third piece= $4x$

$x + 2x + 4x = 175$ inches
$7x = 175$
$x = 25$ inches

Therefore the third piece would be $4(25) = 100$ inches.

28.

a. 27

To solve this problem, we must use the formula for the perimeter of a triangle: $P = a + b + c$. We will use facts from the problem to place in our formula: $P = 64, a = y, b = y + 2$, and $c = 8$.

Next we will plug in the values into the formula: $y + y + 2 + 8 = 64$.

Then combine like terms: $2y + 10 = 64$. After we will subtract 10 from each side to get the variable alone: $2y = 54$

Finally we will divide 2 from each side to find y: $y = 27$ centimeters

29.

a. None, all of these will add up to 9.

From the facts given in the statement and the numbers shown in the answer choices, all of the numbers will add up to 9. 1+3+5=9, 1+8+0=9, and 1+6+2=9.

30.

c. distributive property of multiplication over addition

The number 3 is being distributed to both x and 2 that are added together therefore the distributive property of multiplication over addition.

31.

c. Frank made a mistake in Step 3.

Frank worked the problem correctly until he reached Step 3 where he should have added 28 to each side instead of subtracted since it was -28.

32.

b. This statement is true for the numbers zero and one.

When you square the number zero, it remains a zero. The same is true for the number one, when it is squared it remains a one.

33.

d. 10

Out of all the numbers given 10 is the counterexample to prove Jordan's conclusion false.

$(10)^2 - 5(10) + 10$

$100 - 50 + 10$

60, this answer ends in 0 proving Jordan's conclusion false.

34.

d.

Out of the four graphs, d best represents the equation $y = 4x + 1$ since the graph would need to intersect at 1.

35.

c. $y \geq \frac{1}{3}x - 2$

Out of the four choices given to represent the graph, this equation best fits the graph shown.

36.

a. $y = x + 3$

Out of the four equations given, the equation $y = x + 3$ best represents what is shown on the graph.

37.

b. (6,5)

By using each set of numbers in the problem, the only one that will equal 32 is the set (6,5) with $x = 6$ and $y = 5$

$2(6) + 4(5) = 32$

$12 + 20 = 32$

$32 = 32$

38.

d. (7,-4)

This set of numbers is the solution for the system of equations. While the set (1,8) would fit the first equation, it would not the second.

39.

a. two lines intersecting at one point

The lines have opposite slopes $(-\frac{4}{3}, \frac{4}{3})$ when the equations are put in the $y = mx + b$ form, where m is the slope. Usually when there are opposite slopes, they will intersect at least at one point.

$-4x = 3y + 5$

$y = -\frac{4}{3}x - \frac{5}{3}$

$8x = 6y - 1$

$y = \frac{-8}{-6}x + \frac{1}{6}$

$\frac{4}{3} = \frac{-8}{-6}$ $y = \frac{4}{3}x + \frac{1}{6}$

40.

d. 39

To determine how many quarters there are, we first must make an equation. Using d to represent dimes and q to represent quarters and knowing total amount of the coins is $15.85 and can make this equation: $10d + 25q = 1585$

We also know that the there is a total of 100 coins, we can make this equation: $d = 100 - q$

Since we have a working equation for what d equals, we can plug that equation into our first equation to solve for q: $10(100 - q) + 25q = 1585$

Now solve for q: $1000 - 10q + 25q = 1585$

Combine like terms: $1000 + 15q = 1585$

Get q alone: $15q = 585$

$q = 39$

41.

a. Lines l and q are parallel.

Out of the four statements given, the statement that is true about lines l and q is that they are parallel when shown on a graph since they both have the same slope.

42.

c. $2x^2 + 3x - 8$

When combining the like terms, one must remember to distribute the − to each of the parts of the second equation. This can be done by multiplying it by -1, making the following equation:

$3x^2 + 2x - 6 - x^2 + x - 2$

Then combine like terms: $2x^2 + 3x - 8$

43.

b. $8x^2 + 6$

When attempting to find the missing addends for binomials, you can simply subtract the known addend from the sum.

$6x^2 + 15 - (-2x^2 + 9)$

Multiply the found addend by -1: $6x^2 + 15 + 2x^2 - 9$

Combine like terms to get the answer: $8x^2 + 6$

44.

b. $(5z - 9)(z + 6)$

By simplifying the terms in this equation you will get: $5z^2 + 30z - 9z - 54$

Then combine like terms: $5z^2 + 21z - 54$

45.

d. $4(a - 2b)(a - 5b)$

To get the equation in the problem you must keep the 4 on the outside of the first parenthesized equation for it to be simplified into the form it is shown in the problem.

46.

a. $\dfrac{1}{2x^3}$

With this problem, you first take the four and eight and reduce them since they are both like terms. The four becomes a one and the eight a two giving you: $\frac{x^5}{2x^8}$

Then you simplify the x's: $x^8 - x^5 = x^3$

This exponent remains at the bottom, giving you the solution of: $\frac{1}{2x^3}$

47.

c. $3x^2 + 9x$

When solving this problem, you first must multiply $(x + 3)(3x - 1)$ before adding $(x + 3)$

After multiplying, you get: $x + 3 + 3x^2 - x + 9x - 3$

Combine like terms to receive the answer: $3x^2 + 9x$

48.

b. -4

The equation for this word problem would be $x^2 - x = 20$.

One could use each answer choice in the problem to find the correct solution. The only answer choice that would work is -4.

49.

b. $3x - 6$

Out of the factors shown, $3x - 6$ is the only possible answer since both factors must be a minus to have a negative x and a positive number to show the equation Gavin is factoring.

50.

d. $x^2 + \frac{b}{a}x = -\frac{c}{ax}$

This is Step 2 in completing the square. Choice A forgot the x and x^2, along with choice B and choice C is Step 4 in completing the square.

51.

a. 4,3,5

To have the steps in the correct order, Step 4 needs to be after Step 2 followed by Step 3 and Step 5.

52.

d. $\left\{-\frac{2}{3}\right\}$

There is only one solution in the solution set due to the square root in the quadratic equation becomes zero therefore the plus-minus does not work in this case, as shown in the problem below:

$$x = \frac{-(12) \pm \sqrt{(12)^2 - 4(9)(4)}}{2(9)}$$

$$x = \frac{-12 \pm \sqrt{144 - 144}}{18}$$

$$x = \frac{-12 \pm \sqrt{0}}{18}$$

$$x = \frac{-12 \pm 0}{18}$$

$$x = \frac{6(-2)}{6(3)}$$

$$x = -\frac{2}{3}$$

This is an example of a repeated root which is also seen if you factor out the problem, $9x^2 + 12x + 4 = 0 : (3x+2)(3x+2)$

53.

b. The value of $1^2 - 4(3)(8)$ is negative.

While $\sqrt{-95}$ is not a perfect square, the best reason that there is no real solution to this quadratic equation is $\sqrt{-95}$ is a negative value. This would leave one with the equation: $x = \frac{-1 \pm \sqrt{-95}}{6}$

54.

c. $x = \frac{1 - \sqrt{169}}{14}$

When using the quadratic formula for this equation the -1 that is b would become a positive: $-(-1)$ would be 1. While $\sqrt{169}$ is a positive square root after doing these steps:

$$\sqrt{(-1)^2 - 4(7)(-6)}$$

$$\sqrt{1 - (-168)} = \sqrt{169}$$

The denominator would be 14 since 2*7=14.

55.

b. $y = (x + 3)(x + 4)$

Out of the four answer choices, $y = (x + 3)(x + 4)$ would be the only one that could possibly work. This was derived from the factor of $x^2 + 7x + 12 = 0$ to get this equation or by using the quadratic formula.

56.

b. one

The function $y = x^2 - 8x + 16$ has a double root since its factors are $(x - 4)(x - 4)$. Therefore $x = 4$ only making it a double root.

57.

c. 14 inches

To determine the missing base one must know the formula for solving the area of a triangle which is: $A = \frac{1}{2}bh$

We do not know the base nor height but do have the area along with clues to help us find the base of the triangle. We can substitute the height by making the equation: $h = 2b + 4$ and we know the area is 224 so we can multiply the area by 2 to get the base times height which equals 448. We now can use both of these facts to make an equation to solve for b.

$b(2b + 4) = 448$

Distribute the b: $2b^2 + 4b = 448$

Get all numbers to one side: $2b^2 + 4b - 448 = 0$

Divide by 2: $b^2 + 2b - 224 = 0$

Factor: $(b + 16)(b - 14) = 0$

Therefore $b = -16$ or 14, since a base cannot be negative the base must be 14 inches.

58.

a. $\dfrac{2(x+3)}{x-3}$

When simplifying this equation to its lowest terms, you must factor the problem: $\dfrac{2(x+3)(2x+5)}{(2x+5)(x-3)}$

Then cancel out like terms $(2x + 5)$: leaving you with the answer of: $\dfrac{2(x+3)}{x-3}$

59.

b. $\dfrac{n^2}{3n+2}$

When factoring this problem you must take the n^2 and distribute it to the two numbers in the equation making it: $\dfrac{n^2(3n-2)}{9n^2-4}$

Next we must factor the denominator: $\dfrac{n^2(3n-2)}{(3n+2)(3n-2)}$

Then cancel out like terms, leaving us with the answer of: $\dfrac{n^2}{3n+2}$

60.

a. $\dfrac{(x+2)(x+1)}{3}$

To solve for this problem, you can multiply by flipping the second polynomial fraction. It would look like this: $\dfrac{x^2-4}{x+5} \cdot \dfrac{x^2+6x+5}{3x-6}$

Then you must factor out each polynomial: $\dfrac{(x+2)(x-2)}{x+5} \cdot \dfrac{(x+5)(x+1)}{3(x-2)}$

Cancel out like terms that give you the answer of: $\dfrac{(x+2)(x+1)}{3}$

61.

d. $\dfrac{x+4}{3x-4}$

Since $2x - 1$ is in both the numerator and denominator, it can be divided away leaving $\dfrac{x+4}{3x-4}$

62.

a. $\dfrac{4x+40}{x+3}$

Since 4 and 8 are divisible by 4, they can be reduced to 1 and 2 like this: $\dfrac{2}{x+3} \cdot \dfrac{2x+10}{1} =$

There is nothing else that can be reduced so we can now multiply and get the answer: $\dfrac{4x+40}{x+3}$

63.

b. 3

To do this problem we must make an equation. Make x be the time taken for the 330 miles apart. In x hours, the first car will go 70 mph hours or $70x$. (With distance equaling speed*time)

The second car will go 40 mph or $40x$

Our equation would be: $70x + 40x = 330$

Put like terms together: $110x = 330$

Get x alone to get the answer: $x = 3$

Therefore the two cars at three hours of driving would be 330 miles apart.

64.

c. 70 miles per hour

Letting x be the unknown speed for the last two hours, we can make the following distance equation, based upon distance equaling speed*time: $45(3) + 2x = 55(5)$

Then we can multiply: $135 + 2x = 275$

Combine like terms: $2x = 140$

Get x alone to get the answer: $x = 70$ miles per hour

65.

a. $\dfrac{(x+3y)}{(x-2y)}$

First you must factor the equation: $\dfrac{(x-3y)(x+3y)}{(x-3y)(x-2y)}$

Then cancel out like terms $(x-3y)$ to get your answer: $\dfrac{(x+3y)}{(x-2y)}$

Question Number	Correct Answer	Standard
1	D	1.1
2	B	2.0
3	C	2.0
4	A	12.0
5	B	24.2
6	C	21.0
7	B	17.0
8	A	6.0
9	D	8.0
10	C	9.0
11	D	2.0
12	C	16.0
13	A	15.0
14	B	14.0
15	B	5.0
16	D	11.0
17	D	11.0
18	A	14.0
19	C	6.0
20	D	18.0
21	C	7.0
22	B	7.0
23	A	3.0
24	D	2.0
25	C	3.0
26	C	4.0
27	B	5.0
28	A	5.0
29	A	24.1
30	C	25.1
31	C	25.2
32	B	25.3
33	D	24.3
34	D	6.0
35	C	6.0
36	A	6.0
37	B	7.0
38	A	9.0
39	A	9.0
40	D	9.0
41	A	8.0

42	C	10.0
43	B	10.0
44	B	11.0
45	D	11.0
46	A	10.0
47	C	10.0
48	B	14.0
49	B	14.0
50	D	19.0
51	A	19.0
52	D	20.0
53	B	20.0
54	C	20.0
55	B	21.0
56	B	22.0
57	C	23.0
58	A	12.0
59	B	12.0
60	A	13.0
61	D	13.0
62	D	13.0
63	B	15.0
64	C	15.0
65	A	12.0

Algebra I Solutions: Section 2

1.

b. The Commutative Property of Multiplication.

In this equation it shows that $6(x + 2)$ is equal to $(x + 2)6$ since the two are not being distributed, we can see that the commutative property is being displayed.

2.

d. 12

The cubed square root of 27 is 3 and the square root of 81 is 9. Adding together 3+9 will give you an answer of 12.

3.

c. $x^6 x^4$

The expression $x^6 x^4$ is equivalent to $x^3 x^7$ since when you add the exponents of both expressions together they both will equal 10.

4.

a. $\frac{4}{3}$

To get the reciprocal of a fraction, simply flip the fraction. Therefore the reciprocal of $\frac{3}{4}$ is $\frac{4}{3}$.

5.

a. $-\frac{1}{8}$

The multiplicative inverse is another term for reciprocal. Therefore the reciprocal of -8 is $-\frac{1}{8}$.

6.

a. $x = 1 \ or \ x = -5$

Since 3x+6 equal an absolute value, there is more than one possible answer.
Answer 1:
|3x+6| > 9
3x+6 > 9
Subtract 6 from each side
3x > 3
Divide all sides by 3

x > 1

Answer 2:

3x+6 > -14

Subtract 6 from each side

3x > -15

Divide by 3

x > -5

7.

d. $x = -8$ or $x = 8$

Since 2-x equal an absolute value, there is more than one possible answer.

Answer 1:

-2|2-x| > 12

-2(2-x) > 12

Multiply -2 to each part of the problem

-4+2x > 12

Add 4 to each side

2x > 16

Divide 2 from each side

x > 8

Answer 2:

-2(2+x) > -12

Multiply -2 to each part of the problem

-4-2x > -12

Add 4 to each side

-2x > 16

Divide -2 from each side

x>-8

8.

b. $-5x - 15 = x$

To reach this part of the equation, you first must distribute 3 to the equation $(-5 - x)$. This will give you $-2x - 15 - 3x = x$. Next combine like terms, giving you the answer of $-5x - 15 = x$.

9.

c. $2x = 16$

To simplify this equation you must first distribute 7 on one side and -1 on the other. Remember a negative in front of parenthesis can be equated to a -1.

$7x - 21 - 6x + 4 = 2 - x - 3$

Next combine like terms on each side: $x - 17 = -1 - x$

Then get x to one side to get the equivalent answer: $2x = 16$

10.

c. Step 3

The equation is being solved correctly until Step 3 when 18 is subtracted from each side instead of added. The correct solution for this would be $x = 4$.

11.

d. $x > 11$

To solve this inequality, x must be isolated therefore 4 must be added to each side. Adding 7 and 4 together makes 11, making $x > 11$.

12.

c. 27

The number 27 is divisible by 3 and 9, therefore it disproves the statement made about prime numbers ending in 7.

13.

a. $x = -3$ or $x = 3$

The conclusion is the "then" part of a statement which in this case is $x = -3$ or $x = 3$.

14.

a. 3

One of the numbers that can serve as a counterexample to Taylor's conclusion is 3. By placing 3 into the equation we get: $3^2 + 3(3) - 7$

Then simplifying the problem to get: $9 + 9 - 7$

Combine like terms to receive an answer: 11

With the answer of 11 by using x as 3, we have given a counterexample to disprove Taylor's conclusion.

15.

d. multiplication property of equality

Out of the four properties given, the multiplication property of equality is shown since it allows one to multiply the same quantity by both sides of an equation. This is one of the most commonly used properties to solve an equation.

16.

b. Isaac made a mistake in Step 1.

In solving this equation, Isaac did not distribute the 4 to both the n and 10. It was only distributed to the n. If the equation had been solved correctly it would have looked like this:

$n + 4(n + 10) = 80$

$n + 4n + 40 = 80$

$5n + 40 = 80$

$5n = 80 - 40$

$5n = 40$

$\dfrac{5n}{5} = \dfrac{40}{5}$

$n = 8$

17.

d. This statement is always true.

All numbers, including negative numbers and one, have a reciprocal except for zero. $\dfrac{1}{0}$ is undefined therefore zero does not have a reciprocal.

18.

a. 2

To solve for the y-intercept set $x = 0$ to solve for y.

$5(0) - 4y = -8$

$\dfrac{-4y}{-4} = \dfrac{-8}{-4}$

$y = 2$

19.

c. 3y > 4x-2

Out of the four choices given, 3y > 4y -2 is the best choice for the graph shown. The graph is plotted at two points (-1, -2) and (2, 2). We can plug in these numbers into the equation to see if they are equal.

$3y = 4x - 2$

$3(-2) = 4(-1) - 2$

$-6 = -4 - 2$

$-6 = 6$

or

$3(2) = 4(2) - 2$

$6 = 8 - 2$

$6 = 6$

20.

d.

The line of $y = -\dfrac{1}{2}x + 3$ would go through the points of (0,3) and (6,0) therefore graph d is the best representation of the line $y = -\dfrac{1}{2}x + 3$.

21.

b. $y \leq \dfrac{1}{2}x + 1$

Out of the four equations given, $y \leq \frac{1}{2}x + 1$ is the equation best represented by the graph. The line of the inequality $y \leq \frac{1}{2}x + 1$ goes through the points of (-2,0) and (0,1) which are shown on the graph.

22.

c. $y = 3x - 2$

Out of the four equations shown, $y = 3x - 2$ is the equation that best fits the graph. If we were to find the y-intercept for the equation, we would be able to see where y would go on the graph.

$y = 3x - 2$

$y = 3(0) - 2$

$y = -2$

In this case, $y = -2$ and the points on the graph would be (0,-2) thus what was seen on the graph.

23.

d. $\left(1, -\frac{1}{3}\right)$

Out of the four points given $\left(1, -\frac{1}{3}\right)$ is the only one that can fit into the equation to make it equal as shown below:

$2x - 3y = 3$

$2(1) - 3\left(-\frac{1}{3}\right) = 3$

$2 + 1 = 3$

$3 = 3$

24.

a. $y = -4x + 14$

Out of the four equation given $y = -4x + 14$ is the only equation that can equal the point that is given as shown below:

$y = -4x + 14$

$-6 = -4(5) + 14$

$-6 = -20 + 14$

$-6 = -6$

25.

b. $y = 4x + 11$

To find out which equation was used to generate the table given, you must plug the numbers into the equation. Out of the four given only $y = 4x + 11$ is the one that can equal all four points used in the table.

26.

c. $y = \frac{2}{3}x + 2$

For two lines to be parallel to each other, they both must have the same slope. In this case the slope is $\frac{2}{3}$, therefore $y = \frac{2}{3}x + 2$ is parallel to $y = \frac{2}{3}x + 4$.

27.

d. Lines a and b are perpendicular.

Lines a and b are perpendicular to each other when they are graphed. The lines have the same y-intercepts (-1 and -1 respectfully) and opposite slopes ($\frac{2}{3}$ and $-\frac{3}{2}$ respectfully), therefore the two lines would be perpendicular having a 90° angle when graphed.

28.

b. two parallel lines

Since both equations have the same slope of 2, the lines for the equations would be parallel on a graph.

29.

a. $22.00

Let x be the regular price. A 25% discount would be $0.25x$, therefore the discounted price would be $x - 0.25x$. We will set this to the price that Rex paid allowing it to be y. Now we will solve the problem:

$16.5 = x - 0.25x$

$16.5 = 0.75x$

$x = 22$

30.

b. (3,1)

To find the solution to the system of equations, we must substitute. First we need to make the equation $3x - y = 8$ equal y.

We can do this by subtracting: $y = 3x - 8$

Now we can make the equations equal to each other to solve for x, since both equations now equal y.

$x - 2 = 3x - 8$

We need to get each variable on one side: $2x = 6$

Divide each side by 2 to solve for x: $x = 3$

We have found x, now we can take one of the equations to solve for y.

$y = 3 - 2$

Combine like terms to solve for y: $y = 1$.

31.

d. (2,5)

This system of equations only has one solution set of (2,5). To find the solution to the system of equations, we must substitute. First we need to make the equation $3x + 2y = 16$ equal y.

We can do this by subtracting: $2y = -3x + 16$

Then divide by 2: $y = -\frac{3}{2}x + 8$

Now we can make the equations equal to each other to solve for x, since both equations now equal y.

$2x + 1 = -\frac{3}{2}x + 8$

We need to get each variable on one side: $\frac{7}{2}x = 7$

Multiply each side by the reciprocal of $\frac{7}{2}$ ($\frac{2}{7}$) to solve for x: $x = 2$

We have found x, now we can take one of the equations to solve for y.

$y = 2(2) + 1$

$y = 4 + 1$

Combine like terms to solve for y: $y = 5$.

32.

c. $\frac{1}{3x^5}$

To simplify this equation, you must find the common factor amongst the whole numbers. The whole numbers in the equation are 3 and 9, they both can be divided by 3 to make 3 a 1 and 9 a 3. To reduce the x's, you can subtract the denominator of x's from the numerator as shown: x^{9-4}. This number would be reduced to 5, making it: x^5 and the solution to be: $\frac{1}{3x^5}$

33.

a. $4x^2 + x - 3$

When combining the like terms, one must remember to distribute the – to each of the parts of the second equation. This can be done by multiplying it by -1, making the following equation:

$5x^2 - 4x - 7 - x^2 + 5x - 4$

Then combine like terms: $4x^2 + x - 3$

34.

a. $5x^2 - 6x$

When attempting to find the missing addends for binomials, you can simply subtract the known addend from the sum.

$7x^2 - 9x - 2x^2 - 3x$

Combine like terms to get the answer: $5x^2 - 6x$

35.

d. $3x^2 + 1$

To simplify this equation we first must multiply $(x-1)(3x+2)$ together. This will leave our equation like this: $(x+3) + 3x^2 + 2x - 3x - 2$

Next we must combine like terms to get our answer:

$3x^2 + x + 2x - 3x + 3 - 2$

$3x^2 + 3x - 3x + 1$

After simplifying the x cancels each other out, it leaves us with this expression:

$3x^2 + 1$

36.

b. $w^2 + 6w - 42$

To find the dimensions of the window we must take the facts given and put them into an equation. We know that the length of the window is 6 more feet than its width, we can make the length be represented by $6w$. We also know the area of the window, 42 square feet. We don't know the width, this can be represented by w^2. We can set the problem up similar to a polynomial, w^2 can go first, we can add $6w$ to find the missing dimensions and since the area is known we can subtract this from our equation, making this equation: $w^2 + 6w - 42$.

37.

d. $2(a - 3b)(a + 4b)$

Out of the factored forms shown, the only one that will give you the equation of

$2a^2 + 2ab - 24b^2$ is $2(a - 3b)(a + 4b)$ as shown:

$2(a - 3b)(a + 4b)$

Multiply 2 to $(a - 3b)$: $(2a - 6b)(a + 4b)$

Use the FOIL method to get the equation: $2a^2 + 8ab - 6ab - 24b^2$

Combine like terms to get the final equation: $2a^2 + 2ab - 24b^2$

38.

c. $x - 4$

Out of the factors given, $x - 4$ would be the only one that would make the equation $x^2 - 13x + 36$ with $x - 9$.

39.

b. $(2z - 3)^2$

Only $(2z - 3)^2$ is the complete factorization of the equation $4z^2 - 12z + 9$ as shown:

$(2z - 3)^2$

$(2z - 3)(2z - 3)$

$4z^2 - 6z - 6z + 9$

$4z^2 - 12z + 9$

40.

a. $4(3 + t)(3 - t)$

When completing factorizing the equation $4(3 + t)(3 - t)$, you get $36 - 4t^2$ as shown:

$4(3 + t)(3 - t)$

Distribute the 4 to the first parenthesis: $(12 + 4t)(3 - t)$

Use the FOIL method to get the equation: $36 - 12t + 12t - 4t^2$

Combine like terms to get the final equation: $36 - 4t^2$

41.

b. -9

When -9 is multiplied by -9 it makes a positive 81. 81 subtracted from 9 gives you the answer of 72.

42.

a. 9

Take half of the coefficient of x, 3 and square it. This will give you the number 9. Add 9 to complete the square: $x^2 - 6x + 9 = 24$

43.

b. 2,-10

The simplest way to solve this problem is to factor the equation. First we need to get the equation to equal 0, to do this we subtract 20 from each side to make it:

$x^2 + 8x - 20 = 0$

Next we can factor the equation:

$(x - 2)(x + 10) = 0$

After we have factored the problem, we can solve for x:

$x = 2$

$x = -10$

44.

b. $(x + 3)^2 = 21$

Take half of the x-term (divide it by two) and square it. Add this square to both sides of the equation:

$x^2 + 6x + 9 = 12 + 9$

Convert the left-hand side to squared form. Simplify the right-hand side to get the equation that is part of John's solution:

$(x + 3)^2 = 21$

45.

d. $3x - 5$

When factoring this equation, the factors are: $(4x + 4)$ and $(3x - 5)$. Therefore the solution is $3x - 5$.

46.

a. $ax^2 + bx = -c$

The first step in getting the solution for completing the square is to subtract c from each side, therefore the first step would look like this: $ax^2 + bx = -c$

47.

c. no real solution

The reason that there is no real solution to this quadratic equation is because the square root in the problem $\sqrt{-43}$ is a negative value. When the square root is negative in the quadratic formula, this means there is no real solution as shown in solving of the problem below: $x = \frac{-1 \pm \sqrt{(1)^2 - 4(1)(11)}}{2(1)}$

$$x = \frac{-1 \pm \sqrt{1 - 44}}{2}$$

$$x = \frac{-1 \pm \sqrt{-43}}{2}$$

(no real solution)

48.

d. $\frac{7 - \sqrt{13}}{2}$

Since the numbers remaining from the quadratic equation are not divisible, the solution would remain as it is. The seven would become positive because two negatives turn into a positive. This is shown in the problem worked out below:

$$x = \frac{-(-7) \pm \sqrt{(-7)^2 - 4(1)(9)}}{2(1)}$$

$$x = \frac{7 \pm \sqrt{49 - 36}}{2}$$

$$x = \frac{7 \pm \sqrt{13}}{2}$$

49.

d. $y = (2x - 2)(2x + 4)$

Out of the choices given, these two functions equal 1 and -2 as shown:

$2x - 2 = 0$

$2x = 2$

$x = 1$

$2x + 4 = 0$

$2x = -4$

$x = -2$

50.

b. $x = -1$ and $x = -5$

By looking at the graph you can see that on the x axis when $y = 0$, there are two x-intercepts at -1 and -5. This can also be solved through the quadratic formula or by factoring.

51.

c. two

The equation $y = 3x^2 + 10x + 7$ intersects twice on the x-axis, at -1 and $-\frac{7}{3}$. This can be solved by either using the quadratic formula or by factoring.

52.

b. $\frac{1}{2}$ ft.

The total width will be $x + 11 + x$ or $2x + 11$ since there are two sides of the width that are unknown.

This would be the same for the length that would be written as $x + 15 + x$ or $2x + 15$.

Then this gives us the new area of $(2x + 11)(2x + 15) = 192$

Next we will factor and make the equation equal zero:

$4x^2 + 30x + 22x + 165 = 192$

$4x^2 + 52x - 27 = 0$

Once we have our equation equal to zero, we can now use the quadratic formula:

$x = \dfrac{-52 \pm \sqrt{(52)^2 - 4(4)(-27)}}{2(4)}$

$x = \dfrac{-52 \pm \sqrt{2704 + 432}}{8}$

$$x = \frac{-52 \pm \sqrt{3136}}{8}$$

$$x = \frac{-52 \pm 56}{8}$$

$$x = \frac{-52+56}{8} \text{ or } x = \frac{-52-56}{8}$$

$$x = \frac{4}{8} = \frac{1}{2} \text{ or } x = -\frac{108}{8} = -\frac{27}{2}$$

Since a negative value will not work in this context, it can be ignored. Therefore the width of the wooden floor trim is $\frac{1}{2}$ ft.

53.

a. 37 cm

Using l for length and w for width, we know that the length is 3 more than twice the width so we can say: $l = 2w + 3$

We also know the area is 629, therefore: $lw = 629$

We can plug in $l = 2w + 3$ and first solve for w.

$(2w + 3)w = 629$

$2w^2 + 3w = 629$

$2w^2 + 3w - 629 = 0$

Since our number is too large to factor, we need to use the quadratic formula:

$$w = \frac{-3 \pm \sqrt{(3)^2 - 4(2)(-629)}}{2(2)}$$

$$w = \frac{-3 \pm \sqrt{9 + 5032}}{4}$$

$$w = \frac{-3 \pm \sqrt{5041}}{4}$$

$$w = \frac{-3 \pm 71}{4}$$

$$w = \frac{-3+71}{4} \text{ or } w = \frac{-3-71}{4}$$

$w = \frac{68}{4} = 17$ or $w = -\frac{74}{8}$

Since our width cannot be a negative value, the width must equal 17. Now we can take the 17 we found for the value of w and plug it into our equation to find length.

$l = 2(17) + 3$

$l = 34 + 3$

$l = 37$

The length is 37 cm.

54.

d. $\frac{x-2y}{2y}$

To solve this problem you must first factor: $\frac{(x-2y)(x+3y)}{2y(x+3y)}$

$x + 3y$ are like terms and cancel each other out leaving: $\frac{x-2y}{2y}$

55.

b. $\frac{x+1}{x-5}$

To solve this problem you must first factor: $\frac{(3x+1)(x+1)}{(3x+1)(x-5)}$

$3x + 1$ are like terms and cancel each other out leaving: $\frac{x+1}{x-5}$

56.

c. $\frac{x+3}{2x+7}$

To solve this problem you must first factor: $\frac{(2x-7)(x+3)}{(2x+7)(2x-7)}$

$2x - 7$ are like terms and cancel each other out leaving: $\frac{x+3}{2x+7}$

57.

d. $\frac{4a^2}{a+2}$

To solve this problem you must first factor: $\frac{4a^2(a-2)}{(a+2)(a-2)}$

$a-2$ are like terms and cancel each other out leaving: $\frac{4a^2}{a+2}$

58.

a. $\frac{5(b+1)}{b-4}$

To solve this problem you must first factor: $\frac{5b(b+1)}{(b+1)(b-4)} \cdot \frac{(b+1)(b-1)}{b(b-1)}$

Then cancel out similar factors to find the solution: $\frac{5(b+1)}{(b-4)}$

59.

c. $\frac{3x^2-10x-8}{2x^2+7x-4}$

With this fraction, we are simply multiplying the factors in the numerator and denominator: $\frac{3x^2+2x-12x-8}{2x^2+8x-x-4}$

Then we combine like terms to get our answer: $\frac{3x^2-10x-8}{2x^2+7x-4}$

60.

a. $\frac{(x+3)(x-2)}{3}$

To solve for this problem, you can multiply by flipping the second polynomial fraction. It would look like this: $\frac{x^2+6x+9}{x+2} \cdot \frac{x^2-4}{6x+9}$

Then you must factor out each polynomial: $\frac{(x+3)(x+3)}{x+2} \cdot \frac{(x+2)(x-2)}{3(x+3)}$

Cancel out like terms that give you the answer of: $\frac{(x+3)(x-2)}{3}$

61.

b. $8.42

We know that the total pounds used for this blend is 20 pounds since 6+14=20. We also know the amount that the two coffees are but not the price of it all together, this will be our x. We can find this by multiplying the pounds of coffee bought by its price and add them together to find $20x$:

6(9.40)=56.40

14(8)=112

56.40+112=168.40

Now we need to divide by 20 to found out x: $20x = 168.40$

Our solution is $x = 8.42$. That is the price of the coffee blend together per pound.

62.

d. 33%

We know that the total serving of the breakfast cereal and sugar is 60g. We also know that the sugar percentage is 100% or 1.00 and that the sugar in the cereal is 20% or 0.20. What we do not know is the total percentage of sugar in the breakfast cereal now, this can be x. We can find this percentage by multiplying the grams we do know by the percentage and then divide the two added numbers together by the total amount of grams (60) to get our total percentage of sugar.

10(1)=10

50(0.20)=10

10+10=20

We can now take 20 and divide it into our total grams of breakfast cereal: $\frac{20}{60} = 0.33$ or 33%

63.

c. 300 miles

Using the equation $d = rt$, d will represent the distance in miles from Boston and t will represent the time that the slower train is travelling. We will plug in the rate of the first train as $50t$ to the rate of the second train $75(t - 2)$. Since both equations equal d, we can make them equal to each other to solve for t.

Now we can begin our equation:

$50t = 75(t - 2)$

$50t = 75t - 150$

$-25t = -150$

$t = 6$

Now substitute $t = 6$ into train 1

$50(6) = 300$

The faster train will pass the slower train 300 miles from Boston.

64.

b. $\{(0, 6), (1, 7), (2, 8), (3, 9)\}$

B is the only function given in this question. The domain (x) is different from all of the other domains in the other ordered pairs. The domain cannot be the same and be a function. The range (y) can still be the same and still be a function.

65.

d.

Out of the four graphs shown, only graph d shows all positive y-values on the lines for the equation graphed.

66.

a.

Out of the four graphs shown, graph a is the only one that intersects more than at one point causing them not to be functions of x.

Question Number	Correct Answer	Standard
1	B	1.1
2	D	2.0
3	C	2.0
4	A	2.0
5	B	2.0
6	A	3.0
7	D	3.0
8	B	4.0
9	C	4.0
10	C	5.0
11	D	5.0
12	C	24.1
13	A	24.2
14	A	24.3
15	D	25.1
16	B	25.2
17	D	25.3
18	A	6.0
19	C	6.0
20	D	6.0
21	B	6.0
22	C	6.0
23	D	7.0
24	A	7.0
25	B	7.0
26	C	8.0
27	D	8.0
28	B	9.0
29	A	9.0
30	B	9.0
31	D	9.0
32	C	10.0
33	A	10.0
34	A	10.0
35	D	10.0
36	B	10.0
37	D	11.0
38	C	11.0
39	B	11.0
40	A	11.0
41	B	14.0

42	A	14.0
43	B	14.0
44	B	14.0
45	D	14.0
46	A	19.0
47	C	20.0
48	D	20.0
49	D	21.0
50	B	21.0
51	C	22.0
52	B	23.0
53	A	23.0
54	D	12.0
55	B	12.0
56	C	12.0
57	D	12.0
58	A	13.0
59	C	13.0
60	A	13.0
61	B	15.0
62	D	15.0
63	C	15.0
64	B	16.0
65	D	17.0
66	A	18.0

Made in the USA
Middletown, DE
25 January 2020